MUSINGS ON THE COSMOS

GOING BEYOND THE BOX

JAMES D ROBINSON
An Eocrantis Publication

First published in 2018
by
EOCRANTIS Publishing
10 Bewick Crescent
Newton Aycliffe
Co Durham
DL5 5LQ
England

Printed
by
Orphans (Press) Printers
3 Arrow Close
LeominsterEnterprise Estate
Leominster
Herefordshire
HR6 0LD

ISBN : 978-0-9571378-5-1

CONTENTS

ABOUT THE AUTHOR

James D Robinson was born in Darlington, Co Durham, England, in 1949. Apart from writing, he is also actively engaged with many other interests including : photography, drawing and graphic design, whilst academically speaking, he principally favours cosmology, for which he retains an amateur status.

ACKNOWLEDGEMENTS

Grateful thanks are hereby extended to the following : Dr Mathias Jager and colleagues at the ESA/Hubble Public Information Office, for permission to use the photograph of spiral galaxy NGC 1376, on the cover of this book. Also to : Stephen Robinson and Noel Bennett, who both read the manuscript to check for errors in the text.

Note : The text in this book is presented without indents
by design - not error

OTHER BOOKS by JAMES D ROBINSON

THE AUTHENTICATED METEORIC FALLS of the BRITISH ISLES : Pub 2009
THE PSEUDO METEORIC EVENTS of the BRITISH ISLES : Pub 2010
THIS & THAT : A Photographic Album : Pub 2012
THE DAY THE BRIDGES FELL : Pub 2013
HISTORICA PERSONICA : Photographic Portraits captured at Living History
Events : Pub 2013
PHOTOWORKS - A Multiverse of Images : Pub 2014
THE JAMES D ROBINSON TIME CAPSULE PROJECT (booklet) : Pub 2017
THE POSTER DESIGNS of JAMES D ROBINSON : Pub 2017

~~~~~~~~~~~

His work has also been published in :

The Journal of the Royal Photographic Society
Popular Astronomy magazine
Amateur Photographer magazine
What Digital Camera magazine
North East Life magazine
Skermish magazine
&
The Best of British magazine

Additionally, his photo work is represented in the permanent photo collections of the UK's : National Portrait Gallery, and The Picture Library of the Royal Society, and he is the creator of a time capsule,*which is now stored in the Archive Department of the UK's : Durham County Record Office, where it will remain unopened until the 21st of March 2317, making it a time capsule amongst the thousands which are known, with the 5th longest period of closure on record.

*A short film by Noel Bennett, about the time capsule project, is now available to purchase as a DVD. For further details go to : noelbennett@uwclub.net
~~~~~~~~~~~

INTRODUCTION

Although the book you are about to read pertains to the science of Cosmology, or perhaps it is more correct to say : Cosmogony, since the latter relates more specifically to the study of the 'origin' of the universe, its purpose is not to present an in depth study of Cosmogony, but rather to offer some suggestions as to why one viewpoint on the subject may be more acceptable than another - whilst incorporating the author's own beliefs and speculations.
There are of course a multitude of books about the universe out there, with many covering the subject in much the same way. This present work therefore attempts to introduce the subject matter in a less orthodox manner, in the hope it will bring a fresh perspective to the narrative.

James D Robinson
Newton Aycliffe
Co Durham
England

29th July 2018

*"Whether or not it is clear to you,
no doubt the universe is unfolding as it should"*

MAX EHRMANN
(Deisderata : A Poem for a way of life)

FINITE or INFINITE ?

Towards the end of the 1920's, the American astronomer Edwin Hubble, (1889 - 1953) discovered astronomical evidence that the galaxies in the observable universe were moving apart from each other, and were doing so in such a way that the velocities of their separation was directly proportional to the distances between them.

It was the first indication that the universe was apparently expanding, and once the rate of the expansion was calculated, albeit with some degree of uncertainty, it gave rise to the notion that if the universe was expanding 'now,' then in the past, everything in the universe must have been very much closer together in a hot, dense state,[1] (sometimes referred to as the primordial atom,) which in turn suggested that the universe may in fact have had a beginning, (now thought by scientists to have occurred around 13.8 billion years ago,) with the beginning of the expansion itself being labelled as 'The Big Bang'.

Today, the 'Big Bang' theory is widely accepted as the most likely scenario to explain the possible origin of the universe, with support being found within many areas of the scientific community. It also features prominently in scientific literature where the expansion is often illustrated with the analogy of several balloons marked with a number of dots on each surface to represent galaxies, (see fig opposite) with the distance between the dots increasing across the sequence of images. It is important to note however, that it is not the dots themselves, and hence the galaxies which are moving, it is the space between the dot/galaxies which is 'stretching', and as strange as that might seem, it does seem to account for what astronomers observe when they record the position of galaxies in the real universe over time.

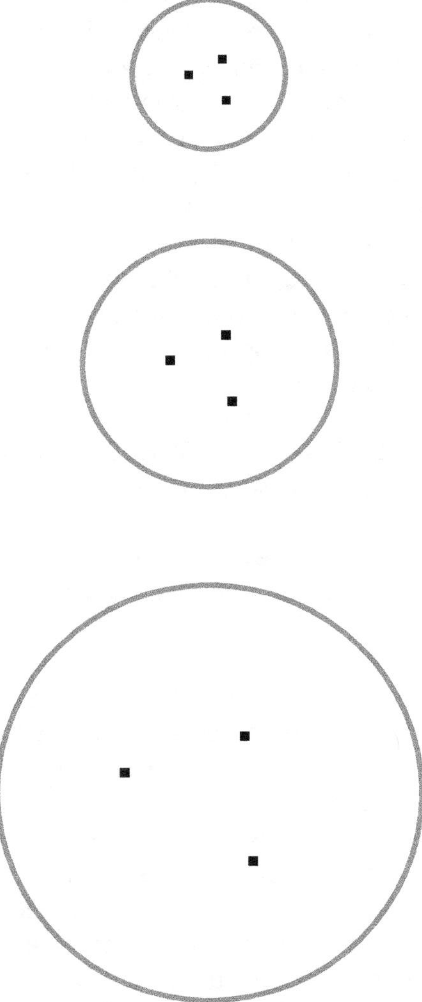

A two dimensional representation of an expanding universe, showing how the space between three galaxies increases over time

On the other hand, there are a number of issues with the Big Bang theory that still need to be resolved because at present, no one can say WHY the expansion began, or how the material that constituted the primordial atom came into being in the first place, other than to voice the vague proposal that it simply popped into existence 'instantaneously' and out of nothingness, which is, to say the least, hardly a satisfactory explanation.

So what does this book have to say on the matter. Did the universe have a beginning, or has it always existed ?

Let's begin by assessing what exactly we mean by the term nothingness, since this would appear to be the most contentious part of the Big Bang theory.

In the context of what we are discussing, by 'nothingness' we mean an ABSOLUTE NOTHINGNESS, and this can be best summed up with the following statement :

Nothingness is less than empty space, because unlike empty space, it has no dimensions.

Alternatively, we can think of it like this : If a circle (as shown opposite) represents the entire universe and all that it contains, and all the contents therein were to suddenly disappear, you would obviously be left with a universe entirely of empty space, i.e, devoid of all matter and all energy. But you would not yet have an absolute 'nothingness', because the empty space of the circle still has a specific dimension, and it still has a dimension in time, equal to the circle's period of existence. It follows therefore, that to attain an 'absolute nothingness', both the dimension of the circle and the circle's lifespan both need to be non-existent, and this is what the Big Bang theory suggests was the state of affairs, BEFORE the universe came into being - bringing time and space with it - somewhere around 13.8 billion years ago. The problem of course, is that for anything to exist 'independently' in reality, it requires dimensions,[2,*] even if they are flexible like the 'forces' in

This circle represents the universe as the home for all that exists
It thus has mass and energy plus dimensions in both space and time

This circle represents the universe when it is devoid of all that exists
except for It's dimensions in both space and time

This image is purely a concept to show that an absolute nothingness
relates to that which is less than empty space, hence the circle is now reduced to dots
whilst the blank space signifies that all things are absent

nature, and since absolute nothingness is (dimensionless') it can only mean there is no such thing as an 'absolute nothingness', other than as a concept. Now, although we might assume that if you once have an absolute nothingness, you will always have an absolute nothingness, there are some who might argue that the universe could still have come into existence by way of 'spontaneous creation'. But how would that work ?. How can you have something come into being when there is nowhere for that to happen, and no time when it could happen ?.

The logical answer is of course that 'no change of ANY sort' is possible if we have absolute nothingness. So the idea of even a spontaneously created univer[*3]se, springing up from nothingness, would seem to be a non-starter. It would therefore be reasonable to accept, based on the above narrative, that if there is no such thing as an absolute nothingness, and given that we know the universe does exist because we are living in it, then there must ALWAYS be a 'SOMETHINGNESS', and thus a universe that has always existed, above and beyond a totally energy free expanse of empty space.

Fortunately for us, a universe comprised of purely empty space is not what we have, which means there has to be something more to the EMPTY SPACE of the universe which will permit itself to somehow transform, in one or more different ways, into the various forms of energy and matter that we know permeate our 'observable universe'.

So how might this new realisation be vindicated?.

Well, as it happens, there is already an established concept that we can look to for support, because it brings us into alignment with one of the features of the Big Bang theory, (or at least with the belief that our observable universe is expanding rapidly) and that is the assumption that what we call 'empty space', is in fact possessed of an energy of its own, intrinsic to its nature, which scientists now refer to, somewhat inappropriately as : VACUUM ENERGY, or alternatively : Zero Point energy.

At present, most descriptions of what this vacuum energy might be, usually lean towards the complex world of quantum physics, that mind-boggling branch of science that deals with the universe on the smallest of scales (much, much smaller than the size of an atom) where strange events are thought to be happening in a way far removed from the physical happenings observed in the macro world.

More specifically, the perception is that 'vacuum energy' is likely the result of 'fluctuations' within the quantum world, so the possibility exists that further studies in quantum physics might eventually provide evidence that supports the notion of an infinite universe of eternal longevity - which would seem a far better option than simply saying that the universe initially came into existence from absolute nothingness. In the meanwhile, until such evidence arrives, or a new theory comes along to conclusively disprove an infinite, eternal[*4] universe, perhaps the best way to summarize all of the above is with the following premise :-

Time, space and the universe, did not have a beginning because that would infer that everything came from absolute nothingness.
But since absolute nothingness equates to a state of
non-existence
by virtue of it having no dimensions
there is no such thing as absolute nothingness, except as a concept.
Therefore, since the universe exists and there is no absolute nothingness, there must always be a somethingness,
therefore the universe has always existed in one form or another,
Therefore, the Big Bang theory, as it currently stands,
is fundamentally flawed.

Of course for anyone who holds a religious belief, God is the creator and ruler of the universe, and nothing more need be said. As for the rest of us, the mysteries of the universe will continue to intrigue, for it is in the nature of humanity to be curious.

FOOTNOTES

*1) The hot dense state of the Primordial Atom is technically referred to (in scientific circles) as a 'Singularity', the definition for which may be expressed as :

*A point in space-time
where matter is infinitely dense
within the smallest volume possible*

where space-time is simply the amalgamation of the 3 dimensions of space with the dimension of time, as envisioned by the German theoretical Physicist, Albert Einstein (1879 - 1955) who saw 'space and time' as a single 'space-time continuum', which extends across the entire universe. It is usually portrayed in reference works as a grid.

*2) A simple way to show that dimensions are needed for something to exist in reality, is as follows : Draw a straight line. The line now has dimensions in both space and time; the later being the period the line remains in view. Now rub out the line and, hey presto, no line and therefore no existence.

3) Since the notion of a physical entity emerging from nothingness may be deemed to have some validity in the quantum world, (given that some sub-atomic particles apparently pop into and out of existence in quick succession), and because 'fluctuations' in the quantum realm are said to be 'unpredictable', 'genuinely spontaneous' and 'intrinsic to nature at it's deepest levels'**, there are some who might view these examples as reason enough to accept that a universe coming from nothingness is a legitimate claim.
However, if you consider the above phenomena whilst holding to

the belief there is no such thing as an absolute nothingness, then the above can be viewed from an entirely different perspective, viz, that instead of the particles fliting in and out of existence, they are instead simply morphing from a state of being detectable to a state of being un-detectable, and thus in reality, never depart the universe, indeed, their very transitory nature might be the cause of, or be influenced by, the fluctuations in the quantum world.

Additionally, it could be argued that : comparing the transitory nature of some sub-atomic particles in the modern day universe, with the supposed birth of the universe from nothingness in the distant past : is entirely incompatible, since the transitory sub-atomic particles of today are immersed in a universe flooded with other entities that may well influence their actions, whereas with the birth of the universe, it is difficult to contemplate what could possibly have initiated ANY KIND of genesis event since, as posited in the Big Bang theory, there was nothing else in existence.

*4) The notion of an eternal universe is very much frowned upon by cosmologists and other academics, because of a notion called Infinite Causal Regression, which posits that since we believe tracing back the cause of a cause cannot go on forever, it implies you will reach a beginning. Thus the universe is seen as having a beginning. But if there was nothing before the beginning (as per the Big Bang theory) then the beginning could not begin. But since the universe exists, the nothing before the beginning must have something. Therefore, there can be no Absolute Nothingness, therefore, the universe (in some form) must exist and be eternal, and the observable universe must be part of a larger universe.

* *Nothingness of Space Could Illuminate the Theory of Everything : Discovery magazine : August 2008*

** *State of the Universe - p5 - New Scientist Supplement : 9 Oct 2004*

THE BEYOND & THE BEYOND THE BEYOND

Now whether you choose to accept or reject the conclusions drawn in the last chapter, there is probably one fact that we can no doubt all agree on, and that is that space, and hence the universe, is awesomely BIG. But just how big is it ? What came before the Big Bang (if indeed there was one ?) and if the observable universe IS expanding, then what is it actually expanding into ?

The truthful answer is that nobody knows for certain, because within any area of research, the search for new knowledge is always going to be compromised by the limitations and the correctness of the knowledge we have already gained, and that being so, our current interpretation of reality is very probably distorted.

But let's take each of the questions above and see what bearing they might have on both the infinite universe option and the Big Bang theory as it stands.

So : WHAT CAME BEFORE THE BIG BANG ?

There was a time, not so very long ago, when virtually every cosmologist, astronomer and physicist, would simply answer : there was nothing before the Big Bang, not even time and space, because of the astronomical evidence discovered by Edwin Hubble back in the late 1920's, and the subsequent belief in an expanding universe. In more recent times however, there has been a growing dissatisfaction with the Big Bang's supposed genesis, and a growing opinion that there must have been something before the Big Bang to initiate that event, even though modern day theoretical physics can trace events back in time to within a fraction of a second immediately AFTER the Big Bang, albeit with a few assumptions along the way.

The reality is however, that at present there is nothing conclusive in the Big Bang theory to say that the universe came from absolute nothingness. It may be heralded by many as a strong possibility, but that still leaves it in the realm of speculation, just as the notion of a universe which is truly infinite in size is currently speculative.

So : HOW BIG IS THE UNIVERSE ?
This question may be answered in three different ways, but only one of them can be expressed with a value in numbers, and that is for the distance between planet Earth and the hypothetical edge of the 'observable universe'.
At present, that distance is estimated to be around 46 billion light years and growing bigger with every second because of the apparent expansion of space, which is increasing at the rate of 70 Km/sec for every MegaParsec of distance, (where 1 Mega Parsec is equal to 3,260,000 light years). The expansion itself is thought to be caused by something cosmologists call : Dark Energy* (see footnote p18.)

For those readers unfamiliar with the astro terminology, a Light year is the distance that light travels in one year through the vacuum of space, and equates to approximately 6 Trillion miles as defined by the International Astronomical Union.
The 46 billion light years measure given above, is calculated by adding the distance light would travel over the estimated age of the observable universe, viz 13.8 billion light years, plus the distance that space has expanded over that same period of time at the above rate of velocity.

Now there may well be an even larger region to the universe beyond what we see in our observable universe, but, sad to say, we may never be able to confirm that supposition as unfortunately - if the expansion of space is true - observing any space beyond our 'observable universe' from our location on Earth, will not be possible.

The reason for this is that as space continues to expand, its most distant region from us will eventually attain a recessional velocity which is greater than the speed of light, (considered to be the speed limit for all physical objects - and light - that may pass THROUGH space, but not for the expansion of space itself) so that any light coming TOWARDS us from any region of space moving faster than the speed of light - will never reach us. It's like walking up an escalator at a certain pace while the escalator is moving downwards at an even faster rate. In such a situation, we would never reach the top of the escalator to see what lay beyond. Likewise, we can only speculate as to what may lay beyond our observable universe, if indeed there is anything to see at all, although currently, no evidence has been forthcoming to suggest any kind of limiting boundary other than the approaching light speed issue as outlined above. Of course if the WHOLE universe IS truly infinite in size, space will continue outwards for ever in all directions - with no overall expansion, even if local regions in the universe are themselves expanding.

Think of the process along the lines of charging a battery. Energy flows from one source to another, but the size of the battery remains the same.

So: IF SPACE IS EXPANDING, WHAT IS IT EXPANDING INTO ? Since, as explained above, we cannot see beyond what we call our observable universe, we cannot say categorically what space is expanding into. Logically, if the universe IS larger than our observable universe, then the space from our observable region

Dark energy is regarded as an unknown form of energy, but one which is thought to act as a repulsive force, thus causing the expansion of the universe.

will most likely simply merge into any further region it encounters, which will no doubt already have its own collection of astronomical objects, (and by 'merge' I mean it will simply dissipate its energy into the new region,) again akin to charging a battery as noted above, even if such may not be an ideal analogy. Alternatively, if the observable universe proves to be all there is, then it's apparent expansion will presumably simply stretch what space already exists to even greater dimensions, and possibly for ever, unless gravity stops the expansion and reverses the process. (see pages 47/48 for more on this aspect.) This would of course effectively mean that the observable universe would be expanding into absolute nothingness, which is in conflict with the main premise of this book, viz : that there is no such thing as absolute nothingness. One could argue therefore, that such a conflict actually lends further support to the notion of an infinite universe, because such a universe would offer a solution not only to what the observable universe is expanding into, but also to what fed the so called Big Bang, to set the whole thing going. But like so many things in life, it all depends on how you wish to interpret the facts.

LIGHT FLIGHT

"OK," you might be saying. "So if we are blocked from seeing beyond the limits of our observable universe, is there anything else that might suggest we live in an even wider universe ?"

Well, let me try and answer that question with a quick overview of our present knowledge about the stars and why we believe them to be truly very far away.

In a nutshell, it all comes down to unlocking the secrets that can be found in starlight, for without that information - which has come about from the hard work of many astronomers over the years - we would probably know very little about the universe at large and all its astounding wonders.

Now, as you might imagine, various techniques have been developed to measure the scale of the universe, so I will begin with a brief look at some of the methods employed, highlighting their limitations whilst revealing their abilities to provide us with a kind of distance ladder, that extends outwards into the deepest reaches of space.

1) THE TRIANGULATION METHOD

This method (sometimes called The Parallax Method) is useful for measuring stellar distances out to around a few hundred lightyears, and is generally looked upon as providing the most accurate results of all the various techniques employed.

A simple example to demonstrate what is meant by Parallax is to close one eye, hold up a finger to cover an object in your home, say a doorknob, after which you open your closed eye and cover your other eye. The result seen is that the doorknob appears to shift position because each eye views the doorknob from a slightly different angle.

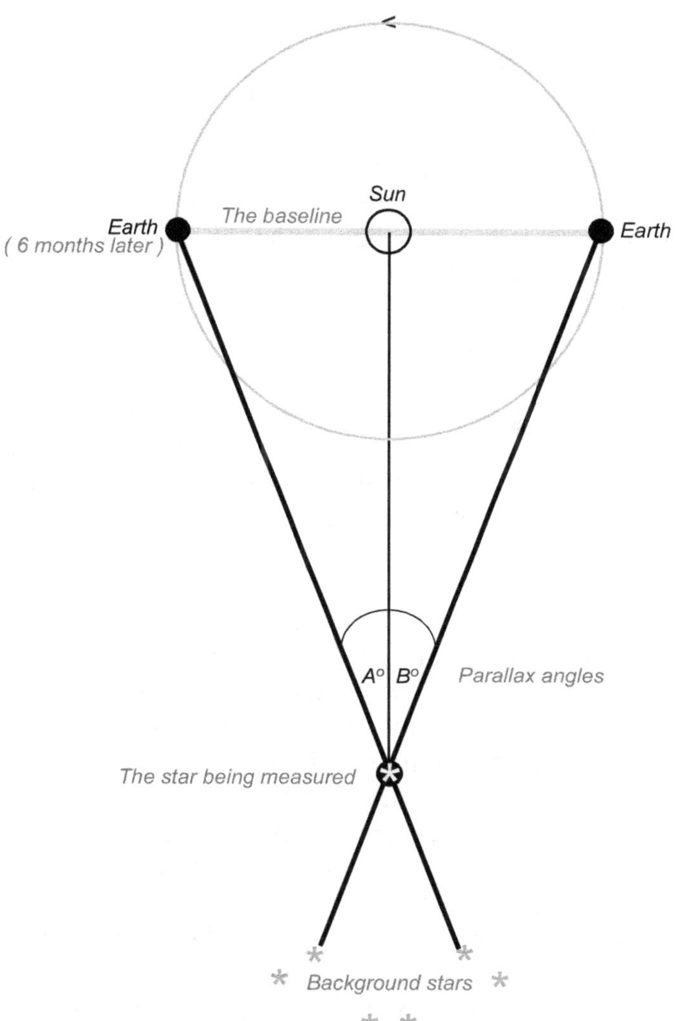

The baseline equals the average
diameter of the Earth's orbit around the
Sun
so once the angle (A + B) divided by 2,
is known
the distance from the star to the Sun
can be calculated using trigonometry

In a similar way, when astronomers want to measure the distance to a nearby star, they can do so by measuring the angle which the chosen star makes, relative to other stars in the background. The measurements are derived from photographs taken 6 months apart when the Earth is on opposite sides of it's orbit around the sun, as this allows the known mean distance of the Earth from the sun to be used as baseline for a right angle triangle (as shown in the illustration on page 21.) Once the angles have been established, the distance of the chosen star from the sun can be determined via trigonometry.

Although the method may appear to be a simple affair, in reality it is not, because the angles involved are exceedingly small and the Earth's atmosphere tends to blur the image of the observed star, which, like all distant stars. is never seen as anything greater than a tiny dot, even in the largest telescopes.

THE VARIABLE STAR METHOD.
This method is useful for estimating stellar distances both within and beyond our own galaxy The Milky Way. It was developed in the opening decades of the 20th century, from work by Henrietta Swan Leavitt who observed that certain types of stars known as Cephid Variables were changing in brightness on a regular basis and over specific amounts of time. The stars themselves, (being very bright and classified as yellow supergiants) were observable both within our own galaxy and in the nearer galaxies that lay beyond, whereafter it was realised that the cephid stars could be used as indicators of stellar distance, if certain measurements were obtained and the Inverse Square Law applied.
In simple terms, this is how the Inverse Square Law works when used in astronomy :
If you have two stars of the same type, they will have the same REAL brightness, even if they appear to be different when viewed from Earth. So if, for eg, you know the distance to the brighter of the two stars, having already determined it's distance by way of

the triangulation method, then the distance to the less bright star can be found when you apply the inverse square law, because light intensity deminishes by 1 over the distance squared.

Here is an example : If the less bright star is found to be only a 1/4 as bright as the brighter star when viewed from Earth, the less bright star will be twice as far away as the brighter star, because one over 2^2 equals 1/4. Likewise, if the less bright star appeared to be only 1/9 times as bright as the brighter star, it would be 3 times further away than the brighter star, because one over the square of 3 equals 1/9, and so on.

To be fair, the actual route that astronomers take to obtain the figures required is a great deal more involved than as above, because Cephid Variable stars come in a variety of types, with their 'period of luminosity' (the increase and decrease in brightness) occuring over different periods of time, so astronomers preferably first need to find Cephid stars of like kind in order to achieve the most accurate distance measurements possible. They also have to take into account how much interstellar dust might lay between the observed star and the Earth. So, not an easy task. However, the technique still has a role to play, especially now that modern technology and the use of space probes is helping astronomers to obtain ever more accurate results.

Other stars known as RR Lyrae Variables (classified as Blue Giants) are also useful for measuring stellar distances, although being less luminous than the much brighter Cephid Variables, the use of RR Lyrae Variables are generally restricted to measuring stellar distances within our own galaxy. As for determining the distances to the more remote galactic objects, astronomers look to observing those exploding stars known as type 1a Super novas, because they also brighten and fade over a certain time frame, which allows their real brightness to be calculated and hence their distance from Earth.

As mentioned earlier in this book, it was knowing the distance to some of the remote galaxies and the realisation that they were all apparently moving away from us, that gave rise to the belief in an expanding universe and eventually to the Big Bang model of a finite universe. There may however, be a possible issue with some of the distance estimates as derived from starlight, which we will now examine, noting other factors on route, with respect to the stars themselves.

GHOSTS IN THE COSMOS
When astronomers study the light from the stars, they invariably do so with the aid of a spectroscope attached to their optical telescopes. The spectroscope converts the incoming light from the star or galaxy under scrutiny into the pattern of a spectra, which is then recorded as a stellar or galactic spectra for future analysis.
Such studies have allowed us to gain an insight into the chemical make up of stars, how the colour of stars relate to surface temperature and mass, how the life span of a star relates to how quickly it consumes its fuel, as well as a number of other factors, all of which have now been gathered into various graphs and diagrams which provide an overall view of how the various types of star evolve over time.
In more specific terms, and by agreement, astronomers usually divide stellar spectra into seven main groups with each group subdivided into ten units. The main groups are listed as : O, B, A, F, G, K, and M, - and are best remembered using the mnemonic : Oh Be A Fine Girl Kiss Me - whilst sometimes the letter 'W' and 'O' are added to the top of the list, with 'N' and 'S' added on to the bottom.
The letters are organised with the hottest stars at the top, where surface temperatures can reach 36,000°C or more; such stars being greenish white or blue in colour, with short lifespans lasting 3 to 10 million years. It then runs down to red stars with surface temps in the range of 2,500°C and lifespans of billions of years.

Now, according to the Big Bang theory, all the stars in the observable universe developed millions of years AFTER the Big Bang event. It follows therefore, that no star or galaxy in the observable universe should have an age of 13.8 billion years or more, and for the nearer stars at least, that seems to be the case. Overall however, there could be some discrepancies and a theoretical possibility that some stellar distance measurements may be erroneous, here's how :

Imagine we have a star with a short lifespan of say, just 20 million years, which is situated in deep space billions of lightyears from Earth. During the time that the star is 'active', it would radiate light in all directions - effectively producing what we might call a lightbubble - which would have a maximum extension from the surface of the star of some 20 million lightyears. Now, even though the star will eventually cease to shine, its light bubble will continue its journey across the universe for as long as it's energy persists or it is in some way impeded.

Now imagine further, that a tiny portion of the lightbubble (in the form of a narrow stream of light) begins to pass through the same region of space where the Earth is located. An astronomer on Earth would see the light from the star, as an image of the star when the star was a certain age, and by studying the spectra of that light the astronomer could estimate the lifespan of the star. The astronomer would also be able to estimate the distance of the light source from Earth, by comparing the absolute magnitude of the light with its apparent brightness as seen from Earth. The trouble is, would the distance estimate be erroneous? because the astronomer would have no knowledge of how far the light bubble had really travelled AFTER the star had ceased to shine, and thus how long ago it was when the parent star actually died. In such a situation the astronomer might well be mislead into thinking the star was just 20 million lightyears or less from Earth, and not the biillions of lightyears which was it's true position at the time it ceased to shine. (see the diagram on page 26)

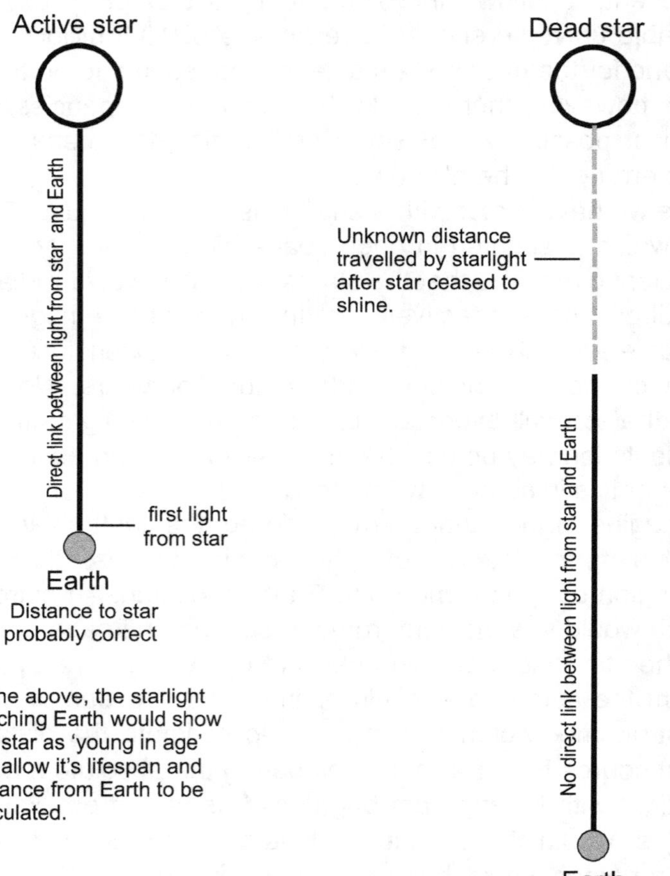

How Stellar distance measurement might be subject to error.

Another way to look at the same issue is as follows : When a distance estimate is made of a star which is still an active stellar object, there is a direct link between the Earth and the star via the light the star radiates. So, if a star with a lifespan of say 10 billion years is situated some 400 million lightyears from Earth, we can picture a stream of light extending continuously through space, from the surface of the star to the astronomer on Earth. However, if the star was situated many billions of lightyears away, and it was no longer shining by the time the lightstream reached the Earth - because the star's lifespan was shorter that the lightstream's 'flight time', then there would no longer be a direct connection between the astronomer on Earth and the star, because a gap would have developed between the surface of the star and the tail end of the lightstream that the parent star radiated into space - as shown in the diagram on page 26.

In such a situation, might it not be therefore, that instead of measuring the distance to the star, we are simply measuring the distance to the remnants of its light ?

Now, admittedly, whilst the figures used above are entirely arbitrary, and the conjecture is speculative, it would appear to have some merit for those stars situated in the more remote regions of space if certain parameters prevailed; parameters which we might define by way of the following expression :

If the lifespan of a star is shorter than the time its light would need to journey to Earth, then all attempts to measure the distance to the star might be subject to error, because part of the starlight's travel time would be an unknown factor.

For eg : if a star (within or beyond the observable universe) has a lifespan of say 5 billion years, and it is situated MORE than 5 billion lightyears from Earth, then any assessment of its distance from Earth might be subject to error. Likewise, the distance to a star with a lifespan of 10 billion years, which is situated MORE than 10 billion lightyears from Earth might also possibly be in error, and so on.

So, what would be the upshot if our stellar distance measurements were, in some cases, short of the mark?.

Well, bearing in mind that stellar distance measurements are useful in judging the distances to galaxies, which in turn have a role to play in estimating the age of our observable universe, incorrect stellar distance measurements would clearly have a major impact on our overall impression of the observable universe, which would no doubt prompt a re-think of the Big Bang theory, since if the most distant stars we see were to be further away than we think, and thus existed at a time older than we think, it would mean that the age of the observable universe itself would potentially be older than we presently think, and that of course would undermine the currently accepted Big Bang theory.

Interestingly, and relating to this very issue, certain astronomical observations of very remote galaxies have come to light in recent years which suggest that their ages may indeed BE much older than we previously thought, because the structure of the galaxies appear to be far more advanced than the current theoretical models of galactic development predict.

To fix numbers to the puzzle, some fully formed galaxies have been found at over 13.1 billion light years distance,* which may have already existed for billions of years, and some concentrations of stars known as globular cluster have been discovered with an age estimated to be greater than 16 billion years, which would clearly make them much older than the present estimated age of the observable universe, assuming the data to be correct.

Additionally, some of the more fully developed galaxies have been found co - existing with more youthful galaxies in the same evolutionary early era of the observable universe, which means we have some galaxies conforming to galactic development predictions and others that do not.

So, could it be that the older galaxies exist, or at least existed at one time, beyond our observable universe, especially if the universe is indeed infinite, since then light from distant sources could have passed our way before the expansion in our region attained light speed ? The search for an answer continues.

THE BIG & THE BIGGER

In addition to the possible issue with regards to stellar distance measurements, there is also another major feature of the Big Bang theory which is generally supported as a sign for a finite universe, which might not be quite what it seems, and the name which that feature has come to be known by is The Cosmic Microwave Background Radiation, or, CMB for short.

It was discovered by accident by Arno Penzias and Robert Wilson, two researchers working for Bell Laboritories back in 1965, revealing itself as a 'hiss' in a large microwave horn they were using at the time for satellite communication experiments.

Since then, it's existence has been confirmed by space probes launched respectively in 1991 and 2003, with the 'hiss' being detected in all regions of the sky.

It has a temperature which measures just a few degrees above absolute zero (which is - 273° C), and it was originally predicted years before, by George Gamow, Ralph Alpher and Robert Herman, as a 'relic radiation' left over from the Big Bang event.

But here's the rub. If, as postulated at the begining of this book, that there must always have been and will always be, some all pervasive form of energy for the universe to exist - because there is no such thing as an absolute nothingness - then that energy, by its very existence, would negate the universe from ever having an absolute zero temperature. So, far from debunking the notion of an infinite universe, the CMB radiation could in fact be an advocate FOR an infinite universe, which has always existed and always will.

"Ah! but," I hear you say. "if the observable universe is expanding as many scientists believe, how can it exist in a universe already infinite in scale and thus at as big a dimension as ever possible?.

Well, if you return to an earlier passage in this book - on page 17 to be exact - you will note that I have already touched upon this issue, though in response to a different question. The jist of it is however, that the answer to your query is all dependent on how you wish to interpret the expansion of space, and how you perceive the Big Bang event itself. But here's a possible option - outside of mainstream orthodoxy.

Since scientists seem to agree that the vacuum of space is actually full of energy, (and an infinite universe must therefore be similarly empowered) then the Big Bang event could have evolved simply as a concentration of such energy (within the infinite universe) which accumulated to such a agree over the passage of time that eventually the intrinsic forces of nature, whatever they were, could no longer contain the concentration of energy, (then high in density and heat) which thus erupted into the general space environment that prevailed around it, dispersing the excess energy, which has been cooling ever since. In other words, the process was entirely within the prevailing laws of nature and did not require some magical output from a 'nothingness' which is beyond logic and reason.

Although the energy dispersal may account for the expansion of space, the CMB radiation and its associated temperature may have more to do with an interaction between the energy released as in the above scenario, coming into contact with the already existing energy of the wider universe - possibly at a boundary layer - and thus acting more akin to the re-charging of a battery, as previously noted on page 17.

If so, the above would negate any conflict with the idea of an expanding region of space existing within an infinite universe, especially if one is also open to the notion that the whole universe is essentially one giant recycling plant. As for the Big Bang being simply a build up of vacuum energy, it should be remembered that the quantum world is often viewed as an environment in constant turmoil, that never comes to total rest, and thus always retains a minimum amount of energy.

It is not totally beyond reason, therefore, to suppose that just occasionally, the energy level in one or more regions of an infinite universe may attain such a highly concentrated form, that eventually it becomes so unstable that the universe releases the source as a superburst of energy, far, far beyond that of supernova explosions, that we would term a Big Bang event.

Now, if all that sounds simplistic you would be right to think so, as there is no certainty that any of the above would, or 'could come to pass in the real universe, especially with respect to the goings on in the quantum world which currently is far from being well understood. Since energy and matter are however, essentially different forms of the same thing, and energetic processes are constantly at work across the entire universe, providing the driving force behind every 'state of change' we observe, perhaps we should not be overly incredulous to the ideas expressed above, and consider instead that they might highlight what may be a universal truth, namely that :

The potential that permits everything that can exist, to exist, is already 'in' the universe, and imparts thereby, of its own accord, in one way or another, an influence on the overall scheme of things, for such is the very nature of the universe, and thus, in consequence, the essence of reality.

LONGEVITY IN THE BLINK OF AN EYE

We will now take a brief detour to contemplate the enigma of time, since an eternal universe equals an infinity of time in combination with space, the two together constituting Einstein's 'space/time continuum'.

It is probably true to say that we frequently use the word 'time' in a variety of expressions, often unaware that we are doing so, when for example, we might say : "It's about time you got here", or, "Don't be late", and yet, if we stop to think about what time really is, we find we are hard pressed to provide a coherent answer.

The problem is, our attempts to describe the real nature of time, often end up being descriptions not about the NATURE of time, but about the PASSAGE of time, which is not the same thing because we are all to well aware of the passage of time, which we recognise as we grow old, and we are all well aware that time can be divided into as many divisions as we may wish, viz seconds, minutes and hours etc, but the nature of time itself remains quite elusive. This chapter will therefore attempt to offer a clearer impression of what time is, and what part it plays in the overall scheme of things. So, onwards and upwards.

It has often been said that time is the fourth dimension, because for any physical object to exist in reality it needs the 3 dimensions of space (length, width and depth) plus a dimension in time, (as noted at the begining of this book.) But time is also a consequence of the occurrence of any event, since no event can take place without some measure of time. We can thus extract from the above the notion that time is a fundamental component of reality, and where events are concerned, it also reveals itself as something that appears to flow in a specific direction, viz from the past to the future, (a fact which is generally referred to as : the arrow of time. But real time has revealed itself to be far more complicated than that, thanks to the work of Albert Einstein, who theorized that the rate at which time passes at a specific reference point is not universal, but is only relative to any other specific reference point, and can be affected by the degree of motion taking place at each reference point, with a higher degree of motion producing a slowing down of the rate at which time passes.

The effect referred to is not noticable in our everyday lives because it only accrues significance at very high velocities, such as might be experienced by an astronaut in a spaceship, moving at close to the speed of light (186,000 miles per second.) Nevertheless, as bizarre as it might sound, scientific experiments with atomic clocks (one at ground level, and another flown in a aeroplane) have shown that the predicated affect (viz that time passes more slowly the faster you move) is indeed real.

The difference in the intensity of gravity at different locations also produces a similar effect, such as for eg, if one clock is sitting on your living room mantelpiece, and another is in a hot air balloon stationed at high altitude above your house. In this situation, the mantelpiece clock would be running at a slightly slower pace than the clock in the balloon, not because of anything to do with the clock's mechanism, but simply because the force of gravity would be stronger in your living room than at the height of the balloon by virtue of it being closer to the Earth's core; a further bizarre feature of nature proven to be real. (see page opposite)

Another aspect of time is the question of whether it flows as an uninterrupted continuity, takes the form of individual pulsations which we might one day be able to measure, or whether it even flows at all. So, let's now examine each concept in turn.

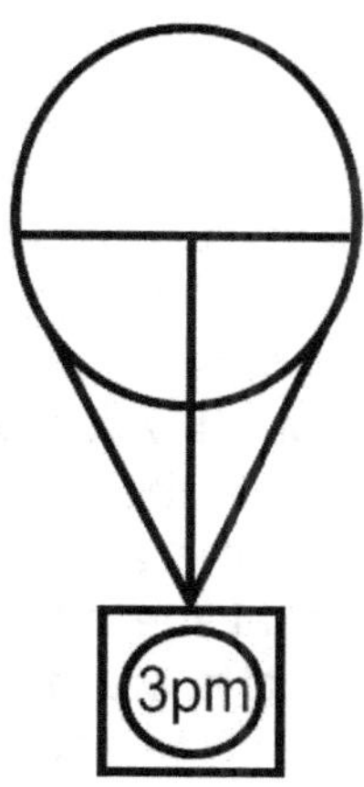

An exaggerated example of how 'gravity' can effect the passage of time for an object on the ground, relative to an object in the sky.
Because the clock is closer to the Earth's core, it sits in a stronger gravity field, so time passes more slowly for the clock, than for the balloon above.
In science, this is termed : Gravitational Time Dilation.

An exaggerted example of how time runs slower for objects in motion, relative to objects standing still. In reality, the difference is not noticeable unless very high velocities are involved.
In science, this is termed : Relativistic Time Dilation.

If time flows as an uninterrupted continuity, and the rate at which it flows, for eg, 1 second per second, is not universal (as suggested by Einstein's work,) it follows that every event in the universe must have an individual timeline, which may or may not match any other timeline, depending on where in space it is occurring, and at what velocity it is occuring. Thus, if multiple events are occurring in the same space environment, permeated with the same level of gravity, and moving at the same velocity, then time will pass at the same rate for each and every event. But if any of the factors are different, then the rate at which time passes will be specific to each location, which means it's rather fortunate for us that such differences of time are only noticable when the differences in motion and gravity strength at any location are extreme.

So, what would we have if time passed as a sequence of rapid pulsations? Well, one obvious thing that springs to mind is that no matter how quickly the pulses of time occurred, you would still have periods between the pulses where time simply did not exist, and that creates a problem, because if you have 'no time', then you also have no reality.

It all boils down again to whether or not you believe in the idea that there is no such thing as an absolute nothingness, because if you DO believe an absolute nothingness CAN exist somewhere in the universe, then you need to resolve how something can exist that has NO dimensions - including no dimension in time. On the other hand, if you believe an absolute nothingness is an impossibility, then the concept of time passing only in pulses would not be part of your philosophy.

The remaining option as mentioned above, about whether time really does flow continuously, is actually quite an interesting topic in it's own right, and carries the narrative back to the idea of individual timelines. Here's why :

If you can accept that the universe is an expanding universe, and really did have a beginning, with nothing existing before it, then

the flow of time would seem self evident as an onward marching phenomenon, subject only to the variations as postulated by Einstein. But if you believe in a universe which is infinite in both space and time, (because there is no such thing as an absolute nothingness,) then OVERALL time could be said NOT to flow, because any passage of time represents an expansion, and any expansion can be traced back to a beginning, which means the universe cannot be infinite in both time and space, and thus we would have a paradox. We can therefore only envisage the passage of time in an infinite universe as a series of non-infinite off shoots - one for every event - which, as individual timelines, arise from an overall, non flowing time pathway, thereby avoiding the paradox, whilst still confirming that all time really is relative. Of course for convenience we still refer to time as something that has a specific value though this is more to do with the biological functions of our brains which recognise the ever changing patterns of reality such as those we experience every day, like rising up in the morning and going to bed at night.
So is time REAL or simply an illusion ?
The truth is no one can tell for certain. Sometimes it seems to flash past. whilst at other times it seems to drag, (another function of the brain no doubt, in reponse to external stimuli, and dependent on how active or inactive we are).
Perhaps therefore, the best way to define the enigma of time is to simply say that it is something that we all 'experience' in our own unique way, and that it is what it is - whatever that is, amen.

THAT THING CALLED ENTROPY

And now back to the main context of this book. Another of the objections to the whole universe having an infinite longevity, is the notion of a process called 'entropy', which scientists have developed to define the way that all real physical processes change over time. It is often cited in reference works as the shift from a state of 'order' to a state of 'disorder', with the central arguments being 1) the amount of entropy never decreases,[*1] and 2) the progress towards disorder cannot be reversed 'naturally', that is to say, without outside assistance, thus a glass jar, which is dropped and broken on the floor, cannot of it's own accord, be put back together again.

In actual fact, it is more correct to describe entropy as 'the dispersal of energy into it's surroundings, when it is free to do so'[*2], as in the scientific arena, entropy relates to 'thermodynamic systems', where such systems are defined as 'specific regions of space in which, at the macroscopic level, one or more thermodynamic processes take place'.

The processes themselves refer to a 'change within a system,' when linked to temperature, pressure, volume, or any transference of heat or energy with its surroundings, whilst the systems aforementioned, are generally referred to as either closed, open, or isolated.

A 'closed' system is defined as : only capable of exchanging energy with it's surroundings, whereas an open system can exchange both energy and matter with it's surroundings.

The third option : an isolated system, is said to be 'in-capable' of exchanging anything with it's surroundings, thus the total universe is viewed as an 'isolated' system, since nothing surrounds it where an exchange of any sort is possible.

The 'observable universe' on the other hand, is said to be an-open system, because it has no discernable boundary, and thus it may interact with whatever may lay beyond the limit of what we are currently able to observe.

Now, taking all the above on board, what the entropy notion boils down to is that over time, the dispersal of energy within the thermodynamic systems will eventually lead to a state of equilibrium across the whole universe, where further exchanges of energy will not be possible, all forms of motion will cease, and all heat sources will thus be extinguished, leaving the universe effectively dead, since even all the particles of the sub atomic world may also decay to nothingness.

It is of course all an assumption based on extrapolating the evidence we have for entropy here on Earth, such as, for example, hot water cooling over time. However, the entropy we see on Earth is occuring in our observable universe (which is NOT in overall equilibrium) so applying the thermodynamic changes to the whole universe as the cause of it attaining an equilibrium, may not be compatible, for as the famous physicist Max Planck[*3] pointed out in his *'Treatise on Thermodymanics,' p103, pub in 1903.*

"We do not say that the 2nd law (of thermodynamics) is applicable to every single detail of a process, for upon closer examination, the matter appears to be this : The entropy, like temperature, pressure and density, cannot be defined as an absolute, continuous quantity, but as a certain average value of a large number of single values. As long therefore, as we regard one or more single values, the entropy cannot be defined any more than the temperature or pressure, and the second law neither applied or proved."

In any event, even if the universe were to attain an overall equilibrium, it will likely remain in that state for evermore (and thus we would still have a universe of sorts) or it would regenerate itself somehow, otherwise it would contravene the first law of thermodynamics which states : Energy can neither be created nor destroyed.

There is also the question of how long attaining a universe wide equilibrium would take, since the 2nd law of thermodynamics offers no indication as to that, and if the universe really is infinite is extension, then it would no doubt take an infinite amount of time to reach equilibrium - which means in effect - equilibrium would never be reached.

It should also be remembered that if, as postulated in this book, there is always an energy of some sort in the vacuum, underlying everything, which we might call the 'base-line energy, which does not itself have to be created because such is the very essence of what we call reality - and thus permits an infinite universe of eternal longevity to exist in the first place - then neither entropy (or anything else) can lead to an absolute nothingness, which suggests the universe has evolved in such a way as to heal itself, after all, we now have stars which are shining, stars which are dead and decaying, and stars which are being born in the vast interstellar clouds of gas and dust which frequent the cosmos.

We thus have hot regions warming cold regions, where, over long periods of time, an equilibrium may be achieved 'locally', but not permanently, since the heat from new stars will continually churn up the space environment, along with any vacuum energy changes or quantum fluctuation which would add to the mix.

There are also, in addition to the above, no doubt many other things in the universe that we don't yet know about, which may well prevent a universe wide 'equilibrium' from coming about, or may develop in the future to prevent such an equilibrium from coming about as the universe continues to evolve.

Yes, its more speculation, but the value of speculation is that it encourages all of us to 'think outside the box', for ultimately it is an initiative which helps to propel progress and avoid academic stagnation.

FOOTNOTES

*1) As described in the 2nd law of thermodynamics.

*2) For a fuller examination on this topic, see the on-line article :
    'Entropy is simple - if we avoid the Briar Path.'
    August 2013
    by : Frank L Lambert, Prof Emeritus (Chemistry)
    Occidental College, Los Angeles.

*3) Max Plank : 1858 - 1947
    German Theoretical Physicist.

PICTURE IT THUS

If the true nature of time gives pause for thought, then so too does the question : Can we objectify the whole universe by giving it a specific shape? or can it only be considered as a multi-dimensional variable, with parameters that may change in scale with the passage of time ? In short, the answer will depend essentially on two factors, viz, whether the universe is infinite or not, and, how much matter it contains.

For those who favour the Big Bang theory, the focus is firmly on the amount of matter in the 'observable universe,' since its density value offers three options as to it's overall shape, each of which is defined in terms of the geometry of it's space/time continuum. This is how the conjecture is posited theoretically :-

If the universe contains a certain amount of matter which cosmologists call The Critical Density - currently thought to be around 3 atoms of hydrogen per cubic meter of space - the geometry of it's space/time will be viewed as 'flat', which simply means that if you drew a triangle on the spacetime continuum the angles of the triangle would add up to 180°, and the universe would be classified as 'open'. If however, the amount of matter in the universe is 'greater' than the critical density, gravity would be strong enough to curve the space/time continuum until it resembled a sphere, whence the universe would be considered as a 'closed structure'. In this condition if you now drew a triangle on the space/time continuum, you would find that the sum of the angles was greater than 180°, whilst if the amount of matter in the universe is less than the critical density, space/time would still be curved, but this time in the shape of a saddle, (aka as a hyperbolic) leaving the universe 'open' with the angles in a triangle totalling less than 180°. (see the page opposite)

POSSIBLE OPTIONS
FOR THE CURVATURE OF SPACE

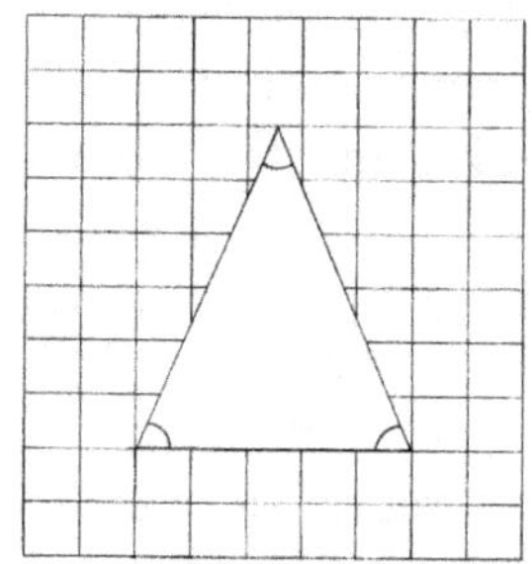

FLAT SPACE - OPEN UNIVERSE - ANGLES = 180°

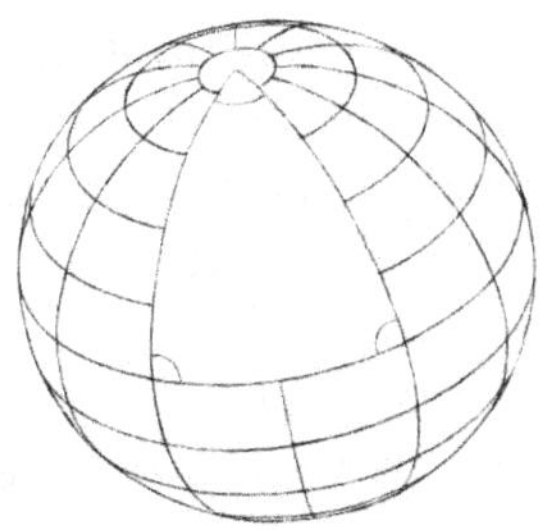

CURVED SPACE - CLOSED UNIVERSE - ANGLES > 180°

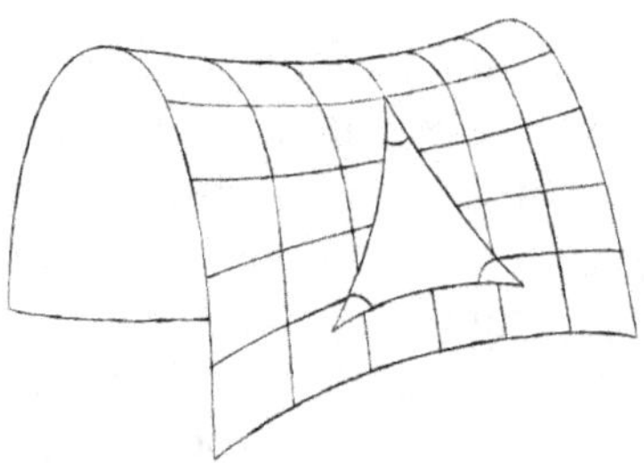

CURVED SPACE - OPEN UNIVERSE - ANGLES < 180°

As things stand, the results of current research suggest that the amount of matter in the observable universe is about equal to the critical density, thus the geometry of its spacetime continuum is thought to be 'flat. Whether the same would hold true in any region of space that might exist beyond our observable universe, is open to question, however, if we accept the assumption that the universe is similar in composition and character in all directions', it would seem probable that the flatness of space will continue as such - and possibly for ever.

So, what's the story for the grander vista, that which cosmologists term the global universe, if it IS infinite ?

Well, the only shape that would seem to be valid would again be a sphere, but this time not one formed by gravity, but simply as a consequence of space reaching out forever in all directions and from whatever starting point you wish to choose. It would thus resemble a sphere of unlimited volume, but one without a unique centre point and a boundary. In other words, it could not be assigned a specific value with respect to its dimension, and every point in space could be equally deemed as the centre of the universe.

That makes it rather difficult to visualise in its completeness, and indeed its complexity is further increased because space in general is full of distortions, regardless of which route you might take to travel within it, since, as Einstein suggested, matter warps space, and the greater the mass of an object, the greater amount that space is warped.

The 'mass' in question belongs of course to the stars, galaxies and planets, or any object of sufficient size and mass to influence the spacetime continuum. Thus, if you were to slice through the universe at any angle of your choosing, whilst its geometry might be classified as 'flat', 'open' or 'closed', its topography would actually be more akin to an undulating landscape, with the degree of undulation being dependent upon the amount of matter situated at each distorted region - see diagram opposite.

In reality, each distortion would entirely encompass the entity causing the distortion, in all three dimensions of space, whilst its position in space would be constantly shifting due to the entities own motion in the space/time continuum.

Now although the distortions themselves are not directly visible evidence of how their gravity shapes the structures within the observable universe as revealed to us in the photographs

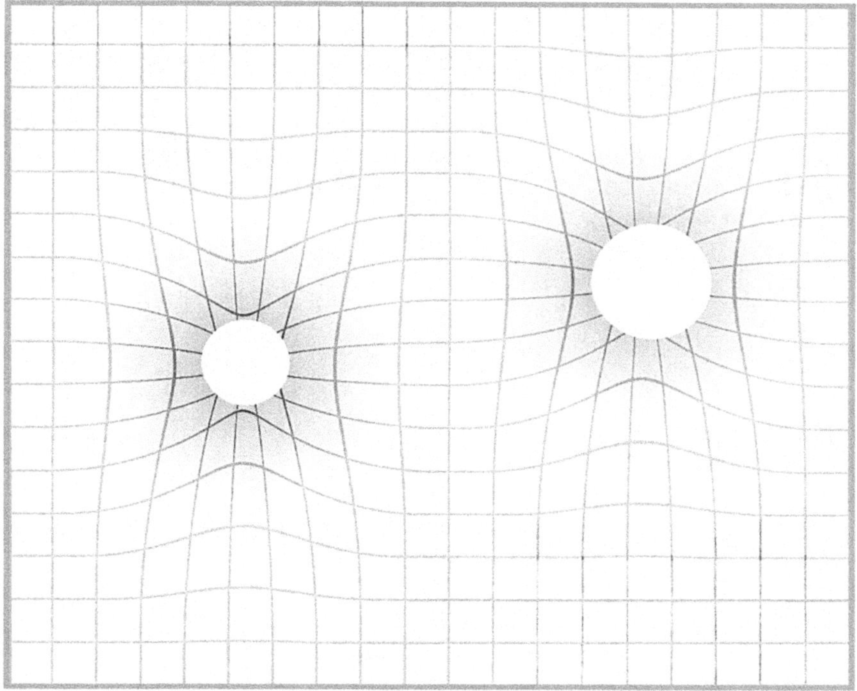

Above. A two dimensional representation of a region of space, (or as Einstein would term it, the space/time continuum) showing how it is warped by the presence of any object with 'mass', such as a star or a planet. The shaded area equates to the object's sphere of gravitational influence, which in theory would extend all the way to infinity, though it's strength, like light intensity, diminishes with distance in accord with the 'inverse square law' as noted on page 16. In more specific terms, the Sun's gravitational influence is said to have a 'noticable effect' on other objects out to a distance of at least a light year, which means it has an influence on those small rocky bodies of the Kuiper Belt, which traverse around the Sun at the very edge of our solar system. As far as we can tell however, none of the nearby stars which lay at a far greater distance, are thought to be in orbit around our Sun.

obtained by astronomers and space probes, which show its grand scale structure to be a mix of high and low concentrations of galactic clusters (and thus regions of varying mass) aligned in long thread-like filaments, or grouped together in sheets inter-spaced with vast areas of near emptiness known as 'voids'.

Frustratingly however, determining if such structures are indeed indicative of a trend that permeates throughout the whole of the universe, be it infinite or otherwise, is something we may never conclusively resolve, for as mentioned earlier in this book (page 12) we appear to be limited on how far we can see into the universe because of the apparent expansion of space within our observable universe. Our best guess is, however, that if there is a wider universe out there, we have nothing concrete to say it will be in any way different to what we observe already, though of course it could be. We can only wait and see which way the pendulum will swing.

ULTIMATE FATE

In addition to affecting the 'shape' of the observable universe, the amount of 'matter' contained in each unit volume of its space also impacts on the amount of energy it contains, and thus on how it operates and evolves, and what its future may be. Unfortunately, the average density value of space is proving difficult to pin down as its calculation is dependent on four other factors* (viz : the Deceleration Parameter, the Cosmological Constant, the Speed of Light and the Hubble Constant) of which only the speed of light is known accurately.

Nevertheless, the general scientific consensus suggests that the 'average density of the observable universe' is roughly equal to the critical density, (and thus as aforementioned - around one atom in every 3 cubic meters of space;) which may not sound very much, but if the universe is infinite in both size and longevity, that value (or whatever the true value is) would still add up to an infinite amount of energy which is thus sufficient enough to account for everything there is, was, or ever will be in the universe, based on the assumption that matter and energy are different forms of the same thing.

On the other hand, for an expanding universe which many believe arose from a moment of creation - as postulated by the Standard Big Bang model - and which therefore is thought to have a finite amount of energy, the average density of space has clear implications for the ultimate fate of the universe, which may play out via one of the following options, relative to the critical value above mentioned, in a manner that echoes the narrative used when discussing 'the shape of the universe'. Thus, if the density of space has a higher value than the critical value, then gravity, it is thought, at some point in time, will halt the apparent expansion of the universe, and reverse its direction leading to a so called Big Crunch, and the likely extinction of everything. But if the density of space is less than the critical value, gravity would be less effective and the universe would expand possibly forever, or until it runs out of energy, leaving thereafter nothing more than an expanse of space which would be cold, dark and utterly lifeless.   As for the critical value itself in relation to the above, this is usually referred to (somewhat vaguely) as 'the minimum density value required to halt the expansion of the universe, but only when applied over an infinite amount of time', which rather leaves the ultimate fate of the universe as an ongoing enigma to which only time has the answer.

Of course if it turns out that the expansion of the observable universe is only a regional event that has little effect on a wider infinite universe, then the Big Bang theory would need to be revised or even abandoned, since its status as the sole origin of everything would clearly be nonsensical.

*DECELERATION PARAMETER : the rate at which the expansion of space is slowing down due to the combined gravitational influence of the galactic clusters.
*COSMOLOGICAL CONSTANT : a supposed force of cosmic repulsion thought to power the expansion of space.
*THE SPEED OF LIGHT : often designated with the symbol 'C' - roughly 186,000 miles per second when passing through a vacuum.
*HUBBLE'S CONSTANT : the rate at which the universe is apparently expanding.

THE IMMORTAL ARENA ?

If there is one area of research which could best help to define reality and in consequence thereby the true nature of the universe, then it may come from a greater knowledge of what are called : The Constants of Nature, a group of parameters viewed as essential to the overall structure of everything, and each with a certain numerical value which is thought to remain 'forever constant' across the eons of time.

Amongst its membership are : the speed of light through a vacuum, the electric charge on an electron, and the gravitational constant, but just why the Constants of Nature have the values they have acquired is presently unknown, which is frustrating for scientists because the values of the constants form the bedrock on which many theories in physics are built, and because if anyone of the constants were to have had a different value, then the universe would have been a very different place. There are three main questions therefore, that Physicists would like to answer viz : 1) Why are the values as they are? 2) Have the values of the constants truly remained the same over the course of time? and 3) If their values have changed, why?.

Now although modern physics has helped to refine the values of the constants, there has been little progress towards their use in creating the hoped for 'Theory of Everything', indeed, if anything, the true nature of the universe has become even more of a muddle as new ideas have entered the scene, bringing forth on the one hand, the concept of a universe imbued with additional dimensions (as promoted in 'M' theory, an off shoot of 'String Theory' : (see footnotes for data on each,) whilst raising, on the other hand, a growing belief that the constants of nature may simply have acquired their traits due to particle interactions way back in the earliest eras of the observable universe.

Researchers have therefore turned towards developing new experiments and theoretical models that may help to clarify if the values of the constants have changed in the real universe compared to the values obtained from lab experiments, with their principal focus now concentrated on assessing a feature known to physicists as 'The Fine Structure Constant'.

The Fine Structure Constant, often referred to simply as 'alpha', relates to the value given to 'the strength of charged particle interactions that occur in outer space, as a consequence of electromagnetic forces.' and it was chosen over reviewing the values of other constants, because any potential change would be the easier to detect (relatively speaking). That said, the Fine Structure Constant is in fact derived from the sum of 4 other constants viz : the speed of light in a vacuum, the value of the electrical charge on an electron, Planck's constant, and a factor called : the vacuum permittivity. (see footnotes for definitions of all) so determining its value still has some issues to overcome, since it has 4 components to measure accurately as opposed to one.

So, in what way might any changes to the Constants of Nature be detected? Well, first up the equipment used must obviously be sensitive enough to detect any change in the value of the constant being studied, whilst remaining free from any factors that might lead to a false result, such as changes to the physical specifications of the research apparatus - and secondly, the source material used to detect any change in the value of a constant should preferably by as old as possible, to cover the chance that some changes might only occur over millions or even billions of years, as compared to any other source used for comparison.

Fortunately, nature itself has provided a number of options which will accommodate the above requirements and they are found when we turn our attention to the stars, and to material from the heavens which has fallen to Earth as meteorites. Here's why :-

In the case of meteorites, they provide material which in most cases is as old as the solar system, thus dating them to around 4.5 billion years of age, which provides a record of what conditions were like in that ancient time. They also provide a chemistry which alters over time through the process of radioactive decay - a change which scientists can measure in accord with certain known laws - which may reveal any changes in the Fine Structure Constant. So far however, only the most minute degree of change has been suggested across the 4.5 billion year time frame, and even that is by no means accepted by all.

The other option relates to those astronomical objects known as Quasars, which are believed to exist in the very remote regions of the observable universe, where, from a relatively small compact source, they shine with a brightness that is equivalent to the total starlight of a galaxy.

It is these characteristics which makes them ideal candidates for checking the value of the Fine Structure Constant, because their remoteness means we are seeing them much further back in time than even the age of meteorites, (due to the speed of light being finite,) and their brightness provides observers with a spectrum in which there is a distinct pattern of dark lines - which are thought to be due to the absorption of certain wavelengths of the Quasar's light by gas clouds which the light encounters whilst on route to Earth. The arrangement of the dark lines gives clues to the value of the Fine Structure Constant, so if the Quasar value turns out to be different to the value of the Fine Structure Constant obtained in the lab, it could potentially indicate that the Fine Structure Constant has NOT remained constant - which would mean a whole re-think of many theories in physics. As things stand however, no conclusive evidence has surfaced to confirm that the Fine Structure Constants value has indeed altered over long periods of time, although some studies[*1] have produced data which leaves food for thought, especially with respect to the speed of light, as any verifiable evidence that the speed of light is NOT constant would throw a spanner in the works for those astro and cosmological theories which use the speed of light to assess the age of the stars and the observable universe - and some recent theoretical studies by scientists in France and Germany might just conceivably show how that could come about.

Put simply, their work involves an assessment of the vacuum of space (which is actually awash with fundamental particles) and the effect it has on the photons of light that pass through it. In other words, it asks : Are there any interactions operating in such a way and to such a degree that it impacts on the value of the speed of light. Here's how they see it :

Since lightwaves have both an electric and magnetic field to which the vacuum of space is, to some degree, resistant, it follows that if the density of the fundamental particles - which may be of various types - is higher or lower in different regions of space, then the interactions with the photons will not be consistent, and hence the speed of light may vary, albeit by a very tiny amount. Unfortunately, whilst further studies are on-going, the experiments needed to validate the concept will have to wait for future technologies because there may be one or more sub-atomic particle types we don't yet know about operating in the vacuum which might have an important role to play in the overall scheme of things, and the only way to check if such unknown particles DO exist is through the use of machines like the Large Hadron Collider working at higher energy levels than in use at present, and such machines are currently unavailable.

Another thing to bear in mind is that there is nothing to say that the interactions referred to above, are not themselves being

*1 *Scientific American magazine, pp 65 - 71 : Inconstant Constants.*
by John D Barrow & John K Webb.

*2 *LiveScience.com : web article : 'Speed of Light may not be constant, Physicists say' by Jesse Emspak, 27th April, 2013.*

influenced by some other actions taking place in a wider universe or indeed in other dimensions that might exist, due to their particles for example, filtering into our observable universe.

Yes - it's more speculation, but then so are other features in modern day cosmology, (the initial Big Bang, Dark energy and the multi-verse to name just three,) all of which creates the problem of choosing which avenues of research to pursue, because something left out (or even overlooked) might very well be an essential part of the overall puzzle, - and that - sadly - is likely to remain the state of play for ever more, because as the astrophysicist Neil deGrasse Tyson once said :

"The universe is under no obligation to make sense to us."

FOOTNOTES

'M' Theory : *A theory in particle physics which postulates that the universe has more dimensions than the 4 we are acquainted with via space and time, and where gravity is unified with the strong and weak nuclear forces. It is also said to encompass all of the various versions of String Theory.*

String Theory : *A theory in physics which postulates that all the elementary particles are nothing more than vibrating 'string-like' entities, in preference to the point-like particles of particle physics.*

The Speed of Light : *A measure of velocity assigned to light and other electro -magnetic radiation, and equal to around 186,000 miles per second .*

The Electric Charge on an Electron : *Equal to an energy value of : 1.60 x 10 to the power of -19 coulombs, according to the Internatiomal System of Units.*

Planck's Constant : *Named after the German Physicist Max Planck. It relates to the amount of energy a Photon carries, relative to the frequency of it's electro-magnetic wave, according to the Simple English Wikipedia.*

Vacuum Permittivity : *said to be a measure which relates to the degree of resistance by the vacuum of space to electric lines of force.*

FINAL SAY

If you have now read this book with an open mind, you will hopefully have come to realise what the question : Did the universe have a beginning, or has it always existed? really boils down to, is : Does everything have a cause and thus a beginning?. Well, here are the options as I see it.
If you believe everything has an underlying cause and thus a beginning, you will most likely be inclined to favour the Big Bang theory. If so, then you need to explain where the 'singularity' which gave rise to the Big Bang event came from, whilst avoiding the problem it poses, since if you have to ask the question, then clearly the 'singularity' cannot be the beginning.
Simply saying it sprang into existence out of an 'absolute nothingness', based on backtracking the expansion of space, and pointing to other evidence which suggests the above might be the case, is an inadequate supposition, and an illogical one at that, since you can never have anything if you have an absolute nothingness. Additionally, a 'moment of creation', would contravene the first law of thermodynamics, which states that : energy can neither be created nor destroyed.
OK, you might say, so what about an eternal universe?, is it not just as illogical to think that something has no beginning, when everything else we no suggests otherwise ?
Well, yes, on the surface it looks that way, but think of it like this : If the universe is eternal, then you can have an infinite succession of events reaching back to a non-definable starting point, which at least sounds more logical than saying something simply popped into existence from an absolute nothingness. In any event, there is another way to decide which is the better option if we apply Occam's Razor, viz which is the simplest solution to resolve most issues, and looked at overall, I would have to suggest that the 'eternal' universe notion is the better, because 1) it upholds the first law of thermodynamics, and 2) it provides solutions for the origin of the 'observable universe', what the observable universe is expanding into, and also for what the underlying nature of reality itself might be, that being the background energy which is believed to permeate all space across the universe.
I am not of course suggesting that the above narrative is necessarily correct, but since we can only work with facts which we think might be correct, I offer the views in this book as reasonable conjectures, and present them here for your own assessment.

REFERENCE SOURCES

Coming of Age in the Milky Way .... *Timothy Ferris*

Before the Big Bang .... *Martin Rees*

The Birth of Time .... *John Gribbin*

Cosmology : A Very Short Introduction .... *Peter Coles*

Quantum Theory : A Very Short Introduction .... *John Polkinghorne*

The Expanding Universe : A Beginner's Guide
to the Big Bang and Beyond : .... *Mark A Garlick*

State of the Universe : A New Scientist supplement : .... *9th Oct 2004*

Origin & Fate of the Universe : Astronomy Magazine special :. *Nov 2004*

Big Bang : A Critical Review   *Ashwini Kumar Lal & Rhawn Joseph*
Journal of Cosmology .... *Volume 6, pp 1533 - 1547, 2010*

Quantum Vacuum Fluctuations
A New Rosetta Stone of Physics ? . ... *Dr H E Puthoff*
        internet article via : www.clas.ufl.edu/anthro/ZPE.html

It's Confirmed   *Stephen Battersby*
Matter is merely vacuum fluctuations .... *Daily News : 20th Nov 2008*
                                                        via the internet

A Brief History of Time .... *Stephen Hawkins*

Many Worlds in One : The search for other Universes .... *Alex Vilenkin*

The Constants of Nature .... *John D Barrow*

An Introduction to Astronomy .... *Robert H Baker*

See also : references noted at the bottom of pages, 15, 29 & 52

FURTHER READING

A selection of other publications offering a more in depth assessment of all things cosmological, whilst generally remaining within the accepted beliefs of mainstream science.

Before The Beginning - Cosmology Expained : G F R Ellis

The Nature of Time & Space : S Hawking & R Penrose

A Short History of the Universe : J Silk

A History of Astronomy & Cosmology : J North

Our Evolving Universe : M S Longair

Astronomy & Astrophysics Encyclopedia : S P Maran

Particle Physics & Inflationary Cosmology : A Linde

Theories of Everything : J D Barrow

About Time : P Davies

The Inflationary Universe : A Guth

Black Holes and the Universe : I D Novikov

Cosmos : C Sagan

Companion to the Cosmos : J Gribbin

The Afterglow of Creation : M Chown

The Stuff of the Universe : J Gribbin & M J Rees

The Accelerating Universe : M Livo

Pathways to the Universe : F Graham-Smith & B Lovell

Ancient Light : A Lightman

www.ingramcontent.com/pod-product-compliance
Lightning Source LLC
Chambersburg PA
CBHW051011050726
47592CB00007B/2795